EMMANUEL JOSEPH

Temperature Treads, Eco-Friendly Fashion for Every Season

First edition

This book was professionally typeset on Reedsy.
Find out more at reedsy.com

Contents

1

Chapter 1: Introduction to Eco-Friendly Fashion

Eco-friendly fashion, also known as sustainable fashion, is a movement that aims to reduce the negative impact of fashion on the environment. It encompasses a range of practices, from using sustainable materials and ethical production processes to promoting recycling and upcycling. The shift towards eco-friendly fashion is not just a trend; it's a necessary change driven by growing awareness of the environmental damage caused by the conventional fashion industry.

One pioneer in this movement is Elizabeth, a former textile engineer. Elizabeth's journey into sustainable fashion began during a business trip to a developing country. There, she witnessed firsthand the pollution caused by textile factories—rivers choked with dye, fields filled with discarded fabrics, and workers exposed to harmful chemicals. Troubled by what she saw, Elizabeth decided to make a change. She left her engineering career and started a sustainable fashion brand. Her designs prioritize organic materials and fair labor practices, proving that fashion can be both beautiful and ethical.

The fashion industry has a long history of fluctuating trends. From the elaborate garments of the Renaissance to the fast fashion of the 21st century, clothing has always been a reflection of society. However, the environmental impact of these changes has often been overlooked. The rise of sustainable fashion represents a shift in priorities, where environmental and social considerations are just as important as aesthetics.

Eco-friendly fashion is not just about the materials used but also about the life cycle of the garments. It involves creating timeless pieces that can be worn for years, rather than disposable items that end up in landfills. The concept of a "capsule wardrobe" has gained popularity, encouraging consumers to invest in a few high-quality, versatile pieces rather than an abundance of cheap, fast fashion items.

For example, Sarah, a fashion blogger, decided to create a capsule wardrobe exclusively with sustainable items. She documented her journey on social media, sharing tips on how to mix and match pieces for different occasions. Her experiment not only made her wardrobe more eco-friendly but also inspired her followers to rethink their own fashion choices.

The growing demand for sustainable fashion has led to innovations in fabric technology. Organic cotton, bamboo, and hemp are just a few of the materials being used to create eco-friendly garments. These materials are not only better for the environment but also often superior in quality. For instance, bamboo fabric is naturally antibacterial and highly absorbent, making it ideal for activewear.

In summary, eco-friendly fashion is a multifaceted movement that seeks to reduce the negative impact of the fashion industry on the environment and society. It involves using sustainable materials, promoting ethical production practices, and encouraging consumers to make more thoughtful choices. As we delve deeper into this book, we will explore the various aspects of eco-friendly fashion and how they can be incorporated into our wardrobes for every season.

2

Chapter 2: The Birth of Sustainable Fabrics

Once upon a time, clothing was primarily made from natural materials like cotton, wool, and silk. These materials were sourced from the environment and, although labor-intensive to produce, were generally biodegradable. However, the Industrial Revolution brought about the rise of synthetic materials such as polyester and nylon. While these materials were cheaper and more durable, they came with a significant environmental cost. The fashion industry's carbon footprint grew, and the issue of textile waste became more pressing.

In response to these challenges, innovators in the fashion industry began to explore sustainable alternatives. Organic cotton emerged as a popular choice, grown without the use of synthetic pesticides and fertilizers. This method of farming not only reduces pollution but also supports biodiversity and healthy ecosystems. The story of Dan and Sarah, founders of a sustainable fashion company, illustrates the potential of organic cotton. Inspired by their love for nature and driven by the desire to make a difference, they started a business that transforms recycled plastic bottles into stylish, durable fabrics. Their journey from concept to reality highlights the innovative spirit within the industry and the possibilities for a greener future in fashion.

Bamboo and hemp are two other sustainable materials that have gained popularity in recent years. Bamboo is known for its rapid growth and minimal need for water, making it an eco-friendly alternative to conventional fabrics. Hemp, on the other hand, is a hardy plant that requires little to no pesticides and enriches the soil it grows in. Both materials are incredibly versatile and can be used to create a wide range of garments, from casual wear to high fashion pieces.

The development of sustainable fabrics has also been driven by advancements in technology. For example, Tencel, a brand of lyocell, is made from sustainably sourced wood pulp and produced in a closed-loop process that recycles water and solvents. This innovative fabric is not only eco-friendly but also exceptionally soft and breathable, making it a popular choice for everything from underwear to evening gowns.

As we move forward, the fashion industry continues to explore new and exciting possibilities for sustainable fabrics. Researchers are experimenting with bio-fabrics made from algae, mushroom leather, and even lab-grown spider silk. These cutting-edge materials have the potential to revolutionize the industry and pave the way for a more sustainable future.

3

Chapter 3: Seasonal Wardrobe Essentials - Spring

Spring is a season of renewal and rebirth, making it the perfect time to refresh your wardrobe with eco-friendly fashion staples. As the weather warms up, light and breathable fabrics become essential. Organic cotton T-shirts, bamboo fiber dresses, and linen pants are just a few of the sustainable options available for spring.

Sarah, a fashion blogger, embarked on a journey to transition her wardrobe to eco-friendly pieces. She faced the challenge of creating a capsule wardrobe for spring using only sustainable materials. Through her blog and social media channels, she shared her experiences and tips for making conscious fashion choices. Her story resonated with many of her followers, who were inspired to make their own wardrobes more sustainable.

One of the key principles of eco-friendly fashion is versatility. By choosing high-quality, timeless pieces, you can create a wardrobe that lasts for years and adapts to different occasions. For example, a classic white organic cotton T-shirt can be dressed up with a blazer and trousers for a business casual look or paired with jeans for a relaxed weekend outfit. Similarly, a bamboo fiber dress can be styled with sandals for a casual day out or dressed up with heels and accessories for a night out.

Accessories are another important aspect of a sustainable spring wardrobe.

Look for items made from recycled or upcycled materials, such as handbags crafted from repurposed fabric scraps or jewelry made from reclaimed metals. These eco-friendly accessories not only add a unique touch to your outfits but also help reduce waste.

By embracing sustainable fashion, you can enjoy the beauty and comfort of spring while minimizing your environmental impact. As we continue our journey through the seasons, let's explore the best eco-friendly options for summer in the next chapter.

4

Chapter 4: Eco-Friendly Fashion for Summer

As the temperature rises, so does our need for comfortable and breathable clothing. Summer fashion trends often lean towards lightweight and vibrant styles, making it the perfect season to showcase eco-friendly materials like linen and Tencel.

Linen, made from the fibers of the flax plant, is a sustainable and versatile fabric that has been used for centuries. It is known for its breathability, durability, and ability to keep you cool in hot weather. Linen dresses, shirts, and pants are summer staples that can be dressed up or down for any occasion. Tencel, made from sustainably sourced wood pulp, is another excellent choice for summer clothing. Its smooth texture, moisture-wicking properties, and biodegradability make it an eco-friendly alternative to synthetic fabrics.

Kevin, a surfer and environmental activist, is passionate about protecting the oceans. He founded a brand that designs eco-friendly swimwear using recycled nylon and polyester from discarded fishing nets and plastic bottles. Kevin's journey began when he witnessed the devastating impact of plastic pollution on marine life during his surfing trips. Determined to make a difference, he started his company with a mission to create stylish, functional, and sustainable swimwear. His story serves as a reminder of the importance of responsible consumer choices and the power of innovative solutions.

Summer accessories, such as hats, sunglasses, and bags, can also be eco-friendly. Look for items made from natural or recycled materials, such as straw hats, sunglasses with frames made from reclaimed wood, and beach bags crafted from upcycled sails. These accessories not only complement your summer outfits but also support sustainable practices.

As you enjoy the sunny days and warm evenings of summer, remember that your fashion choices can have a positive impact on the environment. Embrace eco-friendly materials and support brands that prioritize sustainability. Next, we'll explore the best sustainable fashion options for autumn.

5

Chapter 5: Autumn's Sustainable Wardrobe

With the arrival of autumn, our wardrobes shift to warmer, cozier clothing. Sustainable fashion doesn't have to sacrifice style for comfort. This chapter showcases the best eco-friendly options for autumn, including recycled wool sweaters and organic cotton flannel shirts.

Emma, a teacher and environmental enthusiast, organized a community clothing swap event to promote sustainable fashion. Her initiative encouraged people to bring their gently used clothing and exchange it for new-to-them items. The event was a success, not only reducing textile waste but also fostering a sense of community and shared responsibility. Emma's story demonstrates the power of collective action in driving positive change.

Recycled wool is an excellent choice for autumn clothing. By repurposing discarded wool garments, manufacturers can create new, high-quality pieces while conserving resources and reducing waste. Recycled wool sweaters, scarves, and coats are perfect for staying warm and stylish during the cooler months. Organic cotton flannel shirts, with their soft and cozy texture, are another autumn staple. They can be layered over T-shirts or under jackets for added warmth and versatility.

Layering is key to creating a functional and fashionable autumn wardrobe.

By combining different sustainable pieces, you can adapt to changing temperatures and create a variety of looks. For example, a recycled wool sweater can be paired with organic cotton jeans and a hemp jacket for a casual day out or dressed up with a recycled polyester skirt and eco-friendly boots for a more polished look.

Accessories play an essential role in completing your autumn outfits. Look for items made from natural or recycled materials, such as scarves woven from organic cotton, bags crafted from upcycled leather, and jewelry made from reclaimed metals. These eco-friendly accessories not only add a unique touch to your wardrobe but also support sustainable practices.

As you enjoy the crisp air and changing leaves of autumn, remember that your fashion choices can make a difference. Embrace sustainable materials and support brands that prioritize environmental and ethical considerations. Next, we'll explore eco-friendly fashion options for winter.

6

Chapter 6: Winter Fashion with a Green Touch

Winter brings the need for insulation and warmth. Eco-friendly options for winter wear include recycled down jackets and ethical wool coats. In this chapter, we delve into the stories of companies that have made significant strides in producing sustainable winter clothing.

For example, the founders of a small startup in Norway share their journey of developing a line of eco-friendly thermal wear made from recycled materials. Their dedication to sustainability and innovation showcases the potential for the fashion industry to evolve. The company's thermal wear not only keeps people warm during the harsh Scandinavian winters but also reduces the environmental impact of clothing production.

Recycled down jackets are an excellent choice for winter insulation. By repurposing down from old garments and bedding, manufacturers can create new, high-quality jackets that provide warmth without contributing to waste. Ethical wool coats, made from wool sourced from farms that prioritize animal welfare and sustainable practices, are another great option for staying warm and stylish during the winter months.

Layering is essential for staying comfortable in cold weather. Start with a base layer made from sustainable materials like organic cotton or Tencel, followed by a mid-layer of recycled wool or fleece, and finish with an outer layer of recycled down or ethical wool. This combination of layers will keep you warm and allow you to adjust to varying temperatures throughout the day.

Winter accessories, such as hats, gloves, and scarves, can also be eco-friendly. Look for items made from natural or recycled materials, such as wool hats, gloves made from repurposed fabrics, and scarves crafted from upcycled cashmere. These accessories not only add warmth and style to your winter outfits but also support sustainable practices.

As you embrace the chill of winter, remember that your fashion choices can have a positive impact on the environment. By choosing sustainable materials and supporting eco-friendly brands, you can stay warm and stylish while minimizing your environmental footprint. Next, we'll explore the world of sustainable accessories.

7

Chapter 7: Accessorizing Sustainably

Accessories can make or break an outfit, and there are plenty of eco-friendly options to choose from. From recycled metal jewelry to handbags made from upcycled materials, this chapter explores the wide range of sustainable accessories available.

Meet Lisa, a jewelry designer who started her career by experimenting with discarded materials she found at flea markets and thrift stores. She transformed old coins, bottle caps, and even broken electronics into stunning, one-of-a-kind pieces. Over time, her unique style gained a following, and she launched a successful eco-friendly jewelry brand. Lisa's journey from a hobbyist to a successful entrepreneur is an inspiring tale of creativity and resilience. Her pieces not only look beautiful but also carry stories of transformation and sustainability.

Handbags made from upcycled materials are another fantastic option for eco-conscious fashionistas. Companies like ReBag create stylish bags from reclaimed fabrics and leather, turning what was once waste into fashionable accessories. These bags are not only unique but also help reduce the demand for new materials and decrease the environmental impact of production.

Scarves, hats, and belts can also be made sustainably. Look for items crafted from organic or recycled fibers, such as scarves made from recycled cashmere or hats woven from organic cotton. These accessories add a touch of eco-friendly flair to any outfit while supporting ethical and sustainable practices.

Shoes are another essential part of any wardrobe, and there are plenty of eco-friendly options available. Brands like Allbirds and Veja use sustainable materials like wool, recycled plastic, and natural rubber to create comfortable and stylish footwear. These companies prioritize transparency and ethical production, ensuring that their shoes are kind to both people and the planet.

By choosing sustainable accessories, you can enhance your wardrobe with pieces that are both stylish and environmentally friendly. As we continue our journey through the world of eco-friendly fashion, let's explore the role of technology in driving sustainability in the next chapter.

8

Chapter 8: The Role of Technology in Sustainable Fashion

Innovation and technology have played a significant role in the advancement of sustainable fashion. From 3D printing to biodegradable textiles, technological developments are paving the way for a more sustainable future in the fashion industry.

One example of this is the use of 3D printing to create custom-fit clothing and accessories. By reducing waste and optimizing material usage, 3D printing offers a sustainable alternative to traditional manufacturing methods. The story of a tech-savvy fashion designer who integrates 3D printing into their designs highlights the exciting possibilities that lie ahead for eco-friendly fashion. By creating made-to-order pieces, this designer minimizes waste and ensures a perfect fit for each customer, demonstrating the potential of technology to revolutionize the industry.

Biodegradable textiles are another innovative development in sustainable fashion. Researchers are exploring materials made from algae, mushroom leather, and even lab-grown spider silk. These bio-based materials are designed to break down naturally at the end of their life cycle, reducing the environmental impact of textile waste. The story of a researcher who develops a groundbreaking biodegradable fabric showcases the potential of these materials to transform the fashion industry.

Closed-loop recycling systems are also making waves in sustainable fashion. These systems aim to keep materials in use for as long as possible by recycling and repurposing them at the end of their life cycle. For example, some companies have developed technologies to break down old garments into their basic fibers, which can then be spun into new yarn and woven into new fabrics. This closed-loop approach minimizes waste and conserves resources, making it a key component of a sustainable fashion future.

As technology continues to evolve, the possibilities for sustainable fashion are endless. By embracing innovation and supporting brands that prioritize sustainability, we can help drive positive change in the industry. Next, we'll explore the creative world of upcycling and DIY fashion.

9

Chapter 9: Upcycling and DIY Fashion

U pcycling is a fantastic way to breathe new life into old clothes and reduce waste. This chapter provides practical tips and creative ideas for upcycling and DIY fashion projects.

Consider the story of a group of friends who started a popular YouTube channel dedicated to upcycling tutorials. Their engaging content and passion for sustainability have inspired thousands of viewers to embrace eco-friendly fashion practices. By showcasing simple and creative ways to transform old garments into new, stylish pieces, they have built a community of like-minded individuals who share a commitment to reducing textile waste.

Upcycling can be as simple or as complex as you want it to be. For example, turning an old pair of jeans into a trendy denim skirt requires only basic sewing skills and a few supplies. Similarly, adding patches or embroidery to a plain T-shirt can give it a new lease on life and add a personal touch. More advanced projects might involve combining multiple garments to create a completely new piece, such as transforming a set of old dresses into a patchwork quilted jacket.

In addition to upcycling, DIY fashion allows you to express your creativity while promoting sustainability. Making your own clothes from scratch using eco-friendly materials, such as organic cotton or recycled fabrics, ensures that your wardrobe is both unique and environmentally friendly. There are numerous online resources, including patterns and tutorials, to help you get started with DIY fashion projects.

By embracing upcycling and DIY fashion, you can reduce your environmental impact and create a wardrobe that reflects your personal style. As we continue our exploration of eco-friendly fashion, let's delve into the business side of sustainable fashion in the next chapter.

10

Chapter 10: The Business of Sustainable Fashion

Behind every sustainable fashion brand is a business model designed to minimize environmental impact and promote ethical practices. This chapter delves into the strategies and challenges faced by eco-friendly fashion entrepreneurs.

The story of a husband-and-wife team who launched a successful sustainable fashion brand from their garage serves as a testament to the power of passion and perseverance. Their brand focuses on using organic and recycled materials, ethical production practices, and transparent supply chains. By prioritizing sustainability in every aspect of their business, they have built a loyal customer base and made a positive impact on the environment.

One of the key strategies for building a successful sustainable fashion brand is sourcing materials responsibly. This involves choosing eco-friendly fabrics, such as organic cotton, recycled polyester, and Tencel, and working with suppliers who prioritize environmental and social responsibility. Building strong relationships with suppliers and ensuring transparency in the supply chain are essential components of a sustainable business model.

Another important aspect of sustainable fashion is ethical production. This means ensuring that workers are paid fair wages, work in safe conditions, and are treated with respect. Many sustainable fashion brands partner with factories that adhere to strict ethical standards and invest in programs that support workers' well-being and development.

Marketing and branding also play a crucial role in the success of sustainable fashion brands. By clearly communicating their commitment to sustainability and the positive impact of their products, brands can attract eco-conscious consumers and build a strong brand identity. Social media, storytelling, and collaborations with influencers can help amplify the brand's message and reach a wider audience.

While building a sustainable fashion brand comes with its challenges, the rewards are well worth the effort. By prioritizing sustainability and ethical practices, entrepreneurs can create businesses that make a positive impact on the environment and society. Next, let's explore the power of consumer impact on sustainable fashion.

11

Chapter 11: Consumer Impact on Sustainable Fashion

Consumers play a crucial role in driving the demand for sustainable fashion. This chapter explores the power of consumer choices and how they can influence the industry.

Meet Alex, a college student who started a social media campaign to raise awareness about the environmental impact of fast fashion. His campaign went viral, leading to a significant increase in demand for sustainable fashion products. Alex's story demonstrates the power of individual actions and the influence of social media in driving positive change.

Consumers have the power to shape the fashion industry through their purchasing decisions. By choosing to buy from sustainable brands and supporting ethical practices, consumers can send a strong message to the industry that there is a demand for eco-friendly products. This, in turn, encourages more brands to adopt sustainable practices and develop environmentally friendly products.

In addition to making conscious purchasing decisions, consumers can also reduce their environmental impact by adopting sustainable fashion practices. This includes buying fewer, higher-quality items that are designed to last, repairing and maintaining clothing, and participating in clothing swaps or second-hand shopping. By embracing these practices, consumers can extend the life of their garments and reduce the demand for new production.

Consumer advocacy and activism also play a significant role in promoting sustainable fashion. By raising awareness, educating others, and advocating for policy changes, consumers can help drive systemic change in the fashion industry. Organizations and movements, such as Fashion Revolution, work to promote transparency, sustainability, and ethical practices within the industry, and they rely on the support of engaged consumers to amplify their message.

By making informed choices and advocating for change, consumers have the power to influence the fashion industry and promote a more sustainable future. Next, let's explore the future of sustainable fashion and the emerging trends and innovations shaping the industry.

12

Chapter 12: The Future of Sustainable Fashion

The future of fashion is green, and this chapter explores the emerging trends and innovations that will shape the industry in the coming years. As consumers become more aware of the environmental impact of their choices, the demand for sustainable fashion continues to grow. This shift in consumer behavior drives the industry to innovate and find new ways to reduce its ecological footprint.

One of the most promising developments in sustainable fashion is the rise of bio-based materials. Researchers are exploring alternatives to traditional textiles, such as fabrics made from algae, mushroom leather, and lab-grown spider silk. These materials not only reduce the reliance on fossil fuels but also offer unique properties that can enhance the performance and durability of garments. For instance, spider silk is known for its incredible strength and elasticity, making it an ideal material for sportswear and outdoor gear.

Another exciting trend is the adoption of circular fashion systems. Circular fashion aims to create a closed-loop system where garments are designed, produced, and disposed of in a way that minimizes waste and maximizes resource efficiency. This approach involves using recyclable materials, designing products for longevity, and implementing take-back programs where consumers can return their old clothes for recycling or upcycling.

Brands like Patagonia and Stella McCartney are leading the way in circular fashion, demonstrating that it is possible to create stylish and sustainable clothing.

The role of technology in sustainable fashion cannot be overstated. Advances in 3D printing, AI, and blockchain are revolutionizing the way garments are designed, produced, and distributed. 3D printing allows for on-demand production, reducing waste and overproduction. AI can help designers create more efficient patterns and predict trends, while blockchain ensures transparency and traceability in the supply chain. These technologies are empowering both designers and consumers to make more informed and sustainable choices.

The future of sustainable fashion also involves greater collaboration between stakeholders. Fashion brands, researchers, policymakers, and consumers must work together to drive systemic change. Initiatives like the Fashion Pact and the Global Fashion Agenda bring together industry leaders to set ambitious sustainability targets and share best practices. By fostering collaboration and innovation, the fashion industry can address the environmental and social challenges it faces and pave the way for a more sustainable future.

In conclusion, the future of sustainable fashion is bright, with emerging trends and innovations offering new opportunities for reducing the industry's environmental impact. By embracing bio-based materials, circular fashion systems, and technological advancements, the fashion industry can create a more sustainable and ethical future.

13

Chapter 13: Challenges and Controversies in Sustainable Fashion

While sustainable fashion has made significant strides, it also faces challenges and controversies. This chapter examines the ethical dilemmas, supply chain issues, and greenwashing practices within the industry.

One of the primary challenges is ensuring transparency and traceability in the supply chain. Many fashion brands rely on complex global supply chains, making it difficult to monitor and verify sustainable practices at every stage of production. For example, a brand may source organic cotton, but if the fabric is dyed using harmful chemicals, the overall sustainability of the garment is compromised. The story of a journalist who uncovers the truth behind misleading sustainability claims highlights the importance of transparency and accountability in the industry.

Greenwashing is another significant issue in sustainable fashion. Greenwashing occurs when brands make exaggerated or false claims about the environmental benefits of their products to attract eco-conscious consumers. This practice not only deceives consumers but also undermines the efforts of genuinely sustainable brands. The case of a popular fashion brand accused of greenwashing serves as a cautionary tale, emphasizing the need for rigorous standards and certifications to ensure the authenticity of sustainability claims.

Ethical dilemmas also arise in the context of labor practices and fair wages. While many sustainable fashion brands prioritize ethical production, there are still instances of exploitation and unfair treatment of workers in the supply chain. The story of a factory worker who advocates for better working conditions and fair wages highlights the ongoing struggle for labor rights in the fashion industry.

Despite these challenges, there are efforts to address and overcome them. Organizations like Fashion Revolution and the Clean Clothes Campaign work tirelessly to promote transparency, ethical practices, and workers' rights within the industry. By supporting these initiatives and demanding greater accountability from fashion brands, consumers can play a crucial role in driving positive change.

In summary, while sustainable fashion faces challenges and controversies, addressing these issues is essential for creating a truly sustainable and ethical industry. By promoting transparency, combating greenwashing, and advocating for fair labor practices, we can work towards a fashion industry that prioritizes both people and the planet.

14

Chapter 14: Collaborations and Community Initiatives

Collaborations and community initiatives are essential in promoting sustainable fashion. This chapter showcases inspiring stories of partnerships between fashion brands, NGOs, and local communities.

One such collaboration is between a famous fashion house and a non-profit organization dedicated to environmental conservation. Together, they launched a sustainable fashion line made from recycled ocean plastic. The collection not only raised awareness about marine pollution but also generated funds to support conservation projects. This partnership demonstrates the power of collective efforts in driving change and highlights the potential for fashion to be a force for good.

Community initiatives also play a crucial role in promoting sustainable fashion. For example, a grassroots organization in a small town launched a clothing repair and upcycling workshop. The workshop provided residents with the skills and tools to mend and repurpose their old garments, reducing textile waste and fostering a sense of community. The success of this initiative inspired other communities to start similar programs, creating a network of sustainable fashion advocates.

Educational programs are another important aspect of promoting sustainable fashion. Schools and universities can incorporate sustainability into their curricula, teaching students about the environmental impact of fashion and the importance of ethical practices. The story of a fashion design school that implemented a zero-waste design course highlights the potential for education to drive innovation and inspire the next generation of sustainable fashion designers.

Collaborations between fashion brands and artists can also create unique and impactful projects. For instance, a designer partnered with a renowned artist to create a limited-edition collection using eco-friendly materials. The collection not only showcased the beauty of sustainable fashion but also raised funds for environmental causes. This collaboration demonstrates the potential for creative partnerships to amplify the message of sustainability and reach a broader audience.

In conclusion, collaborations and community initiatives are vital in promoting sustainable fashion. By working together, fashion brands, NGOs, local communities, and educational institutions can drive positive change and create a more sustainable and ethical industry.

15

Chapter 15: Conclusion and Call to Action

In the final chapter, we reflect on the journey of eco-friendly fashion and its impact on the world. Throughout this book, we have explored the various aspects of sustainable fashion, from the development of sustainable fabrics to the role of technology and the power of consumer choices.

The story of a family that transformed their lifestyle to prioritize sustainability serves as a heartwarming reminder of the difference each individual can make. By making conscious choices and supporting sustainable practices, they reduced their environmental footprint and inspired others to do the same.

As we conclude this journey, it's important to remember that sustainable fashion is not just a trend; it's a movement that requires collective action. We all have a role to play in creating a more sustainable and ethical fashion industry. By choosing eco-friendly materials, supporting ethical brands, and adopting sustainable practices, we can make a positive impact on the environment and society.

Let this book be a call to action for each of us to embrace sustainable fashion and contribute to a greener future. Together, we can create a world where fashion is not only beautiful but also kind to the planet and its people.

And so, dear reader, as you continue your journey in the world of fashion, remember the stories, lessons, and inspirations shared in this book. Let them guide you in making thoughtful and responsible choices, and may you find joy in the knowledge that your fashion choices can make a difference.

Description: Temperature Treads: Eco-Friendly Fashion for Every Season embarks on a journey into the heart of sustainable fashion, uncovering the beautiful intersection of style and responsibility. This enlightening book not only delves into the innovative fabrics and ethical practices revolutionizing the industry, but also presents engaging stories of individuals and brands leading the charge towards a greener future.

From spring's refreshing organic cotton T-shirts to winter's cozy recycled wool coats, each season unfolds with eco-friendly wardrobe essentials that marry fashion with sustainability. With practical tips on upcycling, insightful discussions on technology's role in fashion, and inspiring narratives of community-driven initiatives, this book provides a comprehensive guide for the conscious consumer.

Explore the dynamic landscape of sustainable fashion, learn how your choices can make a difference, and be inspired by real-life stories of creativity and commitment. Whether you're a seasoned eco-warrior or new to the concept of green fashion, **Temperature Treads** offers valuable insights and motivation to embrace a wardrobe that is kind to both the planet and its people.

www.ingramcontent.com/pod-product-compliance
Lightning Source LLC
LaVergne TN
LVHW021346200726
843509LV00014B/2690